BEI GRIN MACHT SICH IHR WISSEN BEZAHLT

- Wir veröffentlichen Ihre Hausarbeit,
  Bachelor- und Masterarbeit

- Ihr eigenes eBook und Buch -
  weltweit in allen wichtigen Shops

- Verdienen Sie an jedem Verkauf

Jetzt bei www.GRIN.com hochladen
und kostenlos publizieren

Bibliografische Information der Deutschen Nationalbibliothek:

Die Deutsche Bibliothek verzeichnet diese Publikation in der Deutschen National-
bibliografie; detaillierte bibliografische Daten sind im Internet über http://dnb.d-
nb.de/ abrufbar.

Impressum:

Copyright © 2016 GRIN Verlag, Open Publishing GmbH
Druck und Bindung: Books on Demand GmbH, Norderstedt Germany
ISBN: 9783668359444

Dieses Buch bei GRIN:

http://www.grin.com/de/e-book/346401/lyme-borreliosis-an-emerging-or-re-emer-
ging-disease

Tommy Gong

Lyme borreliosis. An emerging or re-emerging disease?

GRIN Verlag

Lyme Disease

Year 12 Biology ERT

Emerging or Re-Emerging Diseases

Name: Tommy Gong

Table of Contents

Introduction 3

History of Lyme Disease 4

Characteristics of the Pathogen 4

Transmission 5

How *Does Borrelia burgdorferi* Causes Lyme Disease? 5

Stages and Symptoms of Lyme Disease 7

Treatments 8

Statistics and Cases of Lyme Disease 8

Current Management of Lyme Disease 9

Modifications and Improvements to the Management 11

Conclusion 13

Reference 14

Source Analysis 19

Introduction

Diseases are lethal abnormal conditions of the body which stops it from functioning completely or partially (Centers for Disease Control and Prevention, NP). Diseases are separated into two different categories, infectious disease and non-infectious disease. Infectious diseases are diseases that can be transferred to another person through methods of direct or indirect contact. Non infectious diseases, on the other hand, cannot be transferred and can be due to lifestyle and environmental factors or genetics.

Emerging infectious diseases refers to a pathogenic disease that arose recently in a population for the first time (World Health Organization, NP). Re-emerging infectious diseases, however, are diseases that were once a major health problem but has declined due to reasons such as cures being developed. But the pathogen that caused the disease has rearose in a different form that is immune to its original cure (National Institutes of Health US, NP).

Pathogens are biological agents that infects its hosts and causes disruption of its normal physiology through diseases (Science Daily, NP). Pathogens are separated into different categories such as bacteria, viruses, fungi, protozoa, prions and macro parasites (Science Daily, NP). Each of these different types of pathogens have their own ways of transmission, infection, survival in the host and reproduce. In order for a pathogen to be successful, it needs to avoid the host's immune system long enough to infect a new host (Biozone Learning Media Australia, 2014). There are many ways pathogens are transmitted amongst humans such as skin contact, contact of bodily fluids, airborne transmission and vector-borne transmission (Science Daily, NP). Pathogen transmission is efficient and successful in high density populated areas as the pathogens can be easily transmitted from person to person. After pathogens have entered the human body, they begin to multiply. The immune system of the host will respond to the foreign organism by sending out white blood cells and antibodies which functions to rid the pathogens either by directly attacking it or using other mechanisms such as raising the body temperatures which helps rid the viruses as the body becomes a less favorable host (Scientific American, 2005). However, some pathogens are able to evade the immune system and continue to cause diseases. The focus of this research, Lyme disease is an example of this.

History of Lyme Disease

Lyme disease has been around ever since ticks started feeding on human blood. However, the disease was not officially recognized until 1970s where a group of adults and children, from Lyme, Connecticut, started mysteriously suffering from symptoms such as paralysis, rashes and headaches (Bay Area Lyme Foundation, NP). These patients, who were left undiagnosed, started recording notes and researching about their own health problems (Bay Area Foundation, NP). Using the notes that were left by the patients, scientists were eventually able to discover the presence of the infectious bacterial disease that were causing the symptoms and later named it the Lyme disease.

Characteristics of the Pathogen

There is three different types of spiral shaped bacteria, or spirochete that causes Lyme disease (Encyclopedia Britannica, NP). Amongst them, *Borrelia burgdorferi* (Shown in **Figure** 1) is the most common cause of Lyme disease and it mostly affects the North America region. The other two pathogens, *Borrelia afzelii* and *Borrelia garinnii* mainly affects the Europe and Asia regions. *Borrelia burgdorferi* has a length of around 20 μm and a width of only around 1 μm, its growth is ideal in a micro-anaerobic environment at a temperature of 32°C (LymeNet Europe, NP). *Borrelia burgdorferi* is also known for having a flagella located in between the inner and outer membrane of the bacteria which helps it propel into tissues and heavy mucus (MicrobeWiki, 2013).

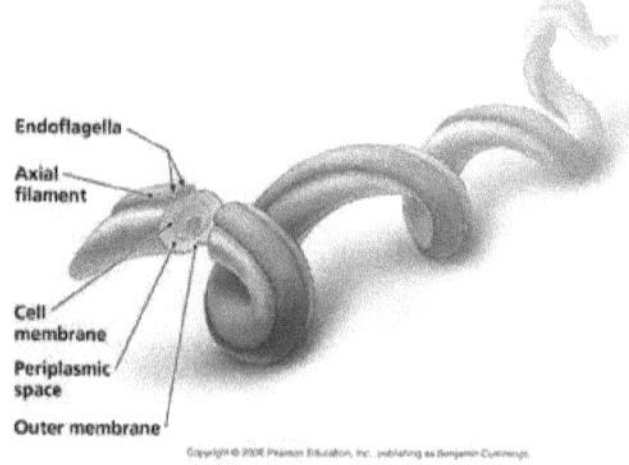

Figure 1. Anatomy of Borrelia burgdorferi (Pearson Education Inc, 2006)

Transmission

Lyme disease is mainly transmitted by ticks of the genus *Ixodes* (Tilly K *et al.*, NP). Ticks require blood meals to complete their life cycles. When a tick that is in its larvae stage has its first few blood meal on small mammals with one of them being the natural reservoir, the pathogen, *Borrelia burgdorferi* from the natural reservoir will then infect the tick and remain in the tick's gut (Centers for Disease Control and Prevention, 2015). The tick will continue to have blood meals from small mammals and simultaneously infecting them with the bacteria, causing them to become natural reservoirs. Once the tick enters its nymph stage, the pathogen migrates to the salivary glands of the tick. These ticks then feeds on larger animals such as humans and will transmit the pathogen during a blood meal. (Bay Area Lyme Foundation, NP). These processes are shown in **Figure** 2.

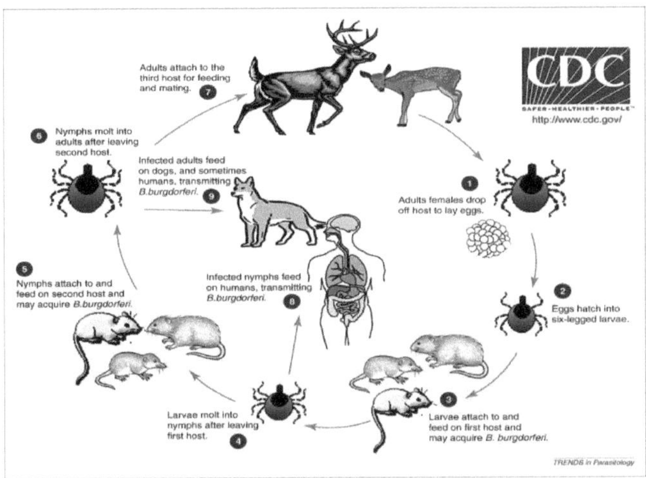

Figure 2. Life Cycle and Transmission of Borrelia burgdorferi in Ticks (Centers for Disease Control and Prevention, NP)

How *Does Borrelia burgdorferi* Causes Lyme Disease?

The immune system protects the body against pathogens and germs. It mainly consists of white blood cells that circulate through the blood stream to detect pathogens. However, *Borrelia burgdorferi* is able to not only evade the immune system's response but is also able to use the immune system against the host. When the pathogen enters the bloodstream, it encounters a dendritic cell which is responsible for alerting the immune system through the presence of a pathogen's antigen (Envita Medical Center, 2008). The bacteria comes into contact of the dendritic cell and rubs its antigen on the cell intentionally. After the dendritic cell has processed the antigen, Helper T cells collects the antigen and passes it to another cell called the Killer T cell. This cell is responsible for directly killing any infected cells (T-cell Modulation Group, NP). The Killer T cells uses the antigen of *Borrelia burgdorferi* to track down the pathogen. Meanwhile, the pathogen spreads its antigen on the surface of a tissue or organ to attract the Killer T cells and the pathogen itself uses its flagella to propel into the tissue or organ (Envita Medical Center, 2008). The Killer T cells then follows the bacteria's antigen to the tissue or organ that has been infected, however, the cell cannot identify and differentiate between the antigen *of Borrelia burgdorferi* and the tissue. Thus, the Killer T cells begins to attack the healthy tissue which causes inflammation in the tissue and destroys the tissue. This is shown in **Figure** 3. The pathogen is also capable of shifting its shape by altering the outer cell wall which "disguises" the pathogen against the immune system (Holtorf Medical Group, NP). As the pathogen spreads through the blood stream, heavy inflammatory responses occur throughout the body which potentially causes organ (including brain) and tissue damages (Holtorf Medical Group, NP).

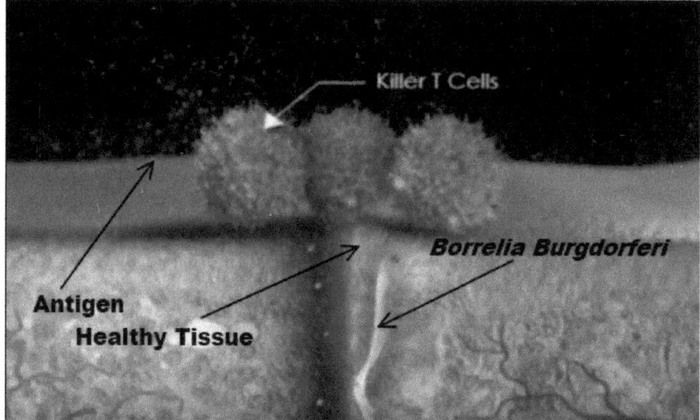

Figure 3. Killer T Cells Destroying the Healthy Tissues (Envita Medical Center, 2008)

Furthermore, the bacteria can also release a neurotoxin named Bacterial Lipoprotein (Envita Medical Center, 2008). This toxin can greatly weaken the immune system and can cause inflammation and damage to the nerve systems of the host which is shown in **Figure 4**. As a result, people who suffer from Lyme disease can have symptoms of memory loss and neurologic pain.

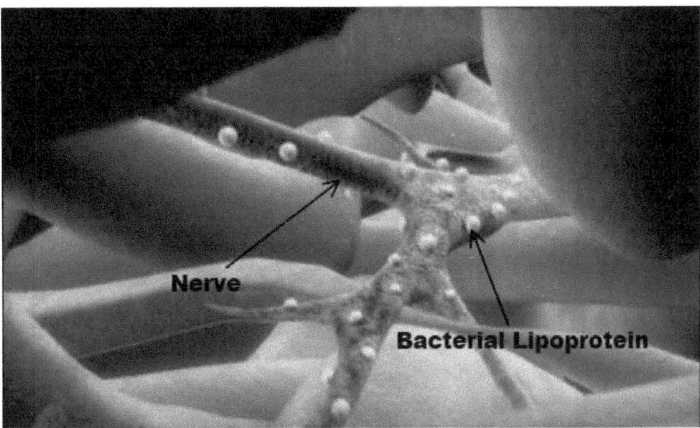

Figure 4. Bacterial Lipoprotein Affecting the Nerves (Envita Medical Center, 2008)

Stages and Symptoms of Lyme Disease

In the first stage, the early localized stage of Lyme disease, the pathogen has just entered the human body. This stage has flu like symptoms and 70 to 80% of patients will also have a red, bull's eye rash at the site of the tick bite (Centers for Disease Control and Prevention, NP). However, sometimes, there may also be no symptoms at all (WebMD, NP). In the second stage, the early disseminated infection, the pathogen has just started spreading around the body and the neurotoxin (Bacterial Lipoprotein) also starts to take effect (Envita Medical Center, 2008). Patients experience memory loss, fainting and rapid heartbeat (WebMD, NP). In the third stage, the late persistent Lyme disease, the bacteria has separated all over the body and affects the joints, nerves and the brain. This stage is the most lethal and symptoms of heart problems, sleep and speaking problems is present. At this stage, the pathogen may also hide in the body and reappear in months or years after the bite from the tick.

Treatments

The most effective treatments currently is the use of antibiotics. At the early stages, treatment is a 14 to 21 day course of either the doxycycline or the cefuroxime antibiotics (Centers for Disease Control and Prevention, NP). All signs and symptoms of infection is usually cleared after the course. However, the treatment has a chance of 14 to 39% failure rate and the symptoms will continue which causes the infection to enter second or third stage (LymeDisease.org, NP). The treatments become more complicated in later stages. At the second stage, depending on the seriousness of the symptoms, patients are either recommended to take oral or intravenous antibiotics and therapy (Hu L, NP). At the third stage of Lyme disease, the patients are recommended to take oral antibiotics for up to 28 days. If symptoms do not get better, it is recommended to take intravenous therapy which in most cases, symptoms will decrease (Hu L, NP).

Statistics and Cases of Lyme Disease

Lyme disease is the most common tick borne disease in the United States. In 2014, 25,359 confirmed cases of Lyme disease was reported (More detailed statistics, **Figure 5**) in the United States (Centers for Disease Control and Prevention, NP). However, the results of studies conducted claims that the actual number of people that are diagnosed with Lyme disease could be up to 300,000 each year (Stricker R, 2014). According to the Lyme Disease Association, a total of 36 deaths were reported from 2002 to 2007, with 2007 having the greatest number of deaths (Refer to **Figure 6**). Furthermore, reported cases of Lyme disease has tripled in 2009 since 1991. Nevertheless, Lyme disease is a continuous disease, it will infect people constantly and did not have any major epidemic or outbreaks.

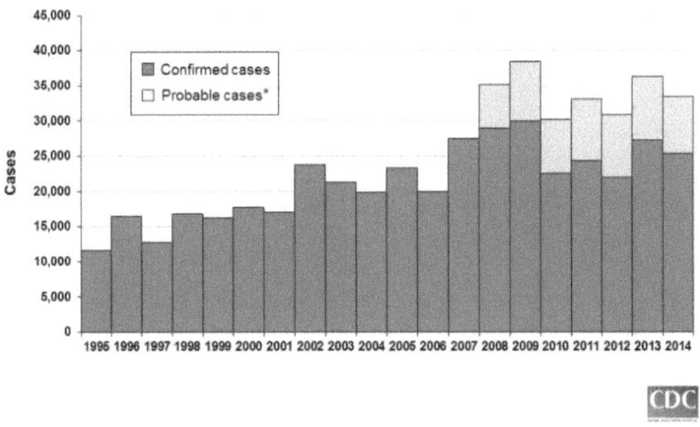

Figure 5. Detailed Statistics on Cases of Lyme Disease From 1995 – 2014 in the United States (Centers for Disease Control and Prevention, NP)

CDC: Deaths from Lyme

Year	2002	2003	2004	2005	2006	2007	TOTAL
Lyme	6	4	6	7	5	8	36

Figure 6. Number of Deaths Due to Lyme Disease in the United States (Centers for Disease Control and Prevention, NP)

Current Management of Lyme Disease

Diagnosing, preventing and curing Lyme disease is difficult due to the mechanisms of the pathogen and its ways of transmission. Furthermore, it is impossible to eradicate the disease unless all of the pathogen's wild reservoirs such as the white footed mouse are eliminated or the *Ixodes* ticks that transmit the pathogen are eradicated completely which is not achievable as it will have a heavy impact on the ecosystems. However, there has been many different approaches and steps taken into trying to prevent Lyme disease. One of the most notable achievement is the successful development of a vaccine to prevent Lyme disease, named LYMErix. The strategy of this vaccine is to vaccinate humans against the protein, Osp A, which is present on the outer membrane of *Borrelia burgdorferi*. This causes the immune system to develop bactericidal antibodies that circulate in the body's blood (Oxford Journals, NP). When the tick ingests the blood during a blood meal, the antibodies enter the tick's guts and effectively bind or neutralize the pathogen present in the tick (Oxford Journals, NP). This vaccine was extremely effective, with a prevention rate of 50 to 100% (Food and Drug Administration, NP). However, due to the public not knowing the dangers of this disease and health professional's lack of knowledge about this disease and vaccine, the purchase of this vaccine was very low and the development was considered unprofitable. Furthermore, hypothesis was made by scientists which claimed that the Osp A protein contained in the vaccine may cause arthritis in people with certain genetic structures which caused continuing pressure placed on the developers of this vaccine by anti-Lyme vaccine groups, the developers soon withdrawn the vaccine from the market, leaving no vaccines available against Lyme disease (Oxford Journals, NP). However, the Food and Drug Administration later claimed that the vaccine was tested negative for the harm that this vaccine has been claimed to be causing (Centers for Disease Control and Prevention, 2015). Nevertheless, the only vaccine for Lyme disease was withdrawn.

Apart from developing vaccines to prevent Lyme disease, many local governments has also tried to warn the public with road signs of heavy tick infested areas. They have also urged the public to adopt tick prevention methods such as wearing long sleeves and using insect repellent when entering tick infested areas (Centers for Disease Control and Prevention, NP). However, this is not effective as some people do not understand the seriousness and the high chances of transmitting diseases like this one from ticks. Furthermore, even with long sleeve clothing, ticks are still able to potentially crawl into clothes through openings. Moreover, most commercial insect repellent uses diethyltoluamide as the main ingredient, but, these kind of insect repellent are not very effective towards ticks as they are against mosquitos (LymeNet Europe, NP). It is also practically impossible to accurately identify the geographical areas of tick infestation which causes many of the road signs to be inaccurate.

Many institutions currently have on going Lyme disease research and companies such as Baxter are developing new vaccines against the pathogen and may be released in the near future (National Institution of Health, NP).

Modifications and Improvements to the Management

Past vaccines has failed due to many factors such as having adverse effects and no market. However, these vaccines are, to some degree, successful in actually working as a prevention method towards Lyme disease. Developers require their vaccines to be profitable or it presents no value to the developers. Thus, to increase the market for Lyme disease vaccines, authorities should begin to firstly recommending this vaccine to people who lives or works in areas of tick infestation. Secondly, trying to raise awareness and educating the general public about the risks and dangers of Lyme disease through sources such as the media.

The next concern is the vaccine itself. Specific requirements must be met for the new vaccine to be successful in the prevention of Lyme disease. Firstly, the vaccine must provide protection for all *Borrelia* bacteria as *Borrelia burgderforgi* is not the only bacteria from that genus that causes Lyme disease. Secondly, the new vaccine must be also efficient and safe for the use of children as children between 5 and 10 years old has more reported cases then all other age groups which is shown by Figure 7. Thus, the use of this vaccine in children will be essential (Centers for Disease Control and Prevention, 2015). Thirdly, the vaccine would require at least an efficiency of 80% otherwise the vaccine would be insufficient in trying to protect the body against Lyme pathogens (Centers for Disease Control and Prevention, 2015). Fourthly, the vaccine needs to have prolonged efficiency or the vaccine will lose its effect after a short amount of time and the patients may need to revaccinate which could cause some patients to choose to not use a vaccine. Lastly, and most importantly, the vaccine must rid most after or side effects which was also one of the main reason that LYMErix was withdrawn.

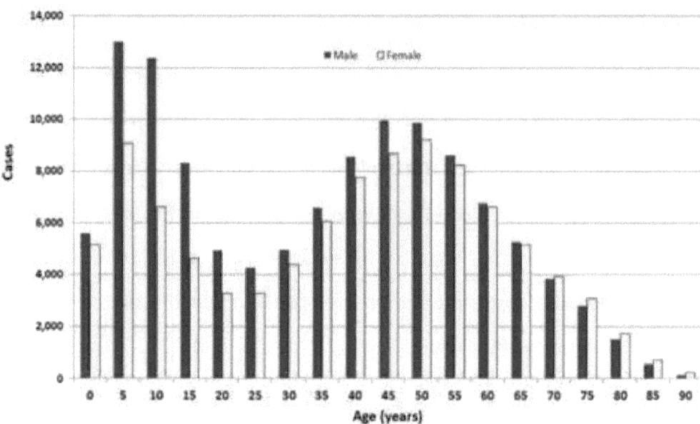

Figure 7. Cases of Lyme disease Based on Age and Sex in the United States (Centers for Disease Control and Prevention, NP)

Fundamentally, more research and development must be put into effectively preventing this disease from infecting more people. However, the research will require great amounts of money. Thus, governments should start considering increasing their funds towards Lyme disease research. The current yearly fund for tick borne diseases is only 74 million and Lyme disease research is only one part of it (National Academy Press, 2011). Yet the funding CDC received from the US government for the use of Ebola responses and prevention had a value of 1.771 billion dollars (Centers for Disease Control and Prevention, NP). However, Lyme disease had around 3,000 more confirmed cases than Ebola. Furthermore, the funding should not only be towards the research of Lyme disease but also contribute to educating the general public about the dangers of Lyme disease and the promotion of new vaccines.

One alternative method that is currently under consideration is to use the chemical named Permethrin which is a neurotoxin that is able to kill ticks when they come into contact (Leprince DJ, NP). This chemical is sprayed onto cotton balls and are distributed in different habitats. Small mammals (potential natural reservoirs) could collect these cotton balls and use it as nesting materials (Leprince DJ, NP). The chemical could then kill any ticks that tries to feed on the small mammals and systematically transmit pathogens to the small mammals (LymeNet Europe, NP). However, this method would be ineffective as the chances of these cotton balls being collected is very low unless a lot of it was distributed. The ticks would also be able to still reach the small mammals as this chemical must come into contact to take effect. Furthermore, the chemical only lasts up to a week and thus cotton balls would need to be continuously manufactured and distributed which makes this method inefficient.

A more feasible method could be to develop a vaccine for small mammals (potential natural reservoirs) (Kurtenbach K *et al.*, NP). The strategy for this vaccine could be similar to LYMErix. This vaccine could be able to potentially deactivate or neutralize the pathogen whilst being transmitted from the ticks. The vaccine can be developed into a form of bait and placed around the habitats of small mammals. If this vaccine is successfully developed and distributed, this method could greatly decrease the population of natural reservoirs, infected ticks and *Borrelia* bacteria. And thus, reducing the number of people being infected. This vaccine must be able to firstly, deactivate or kill the bacteria whilst it is still in the tick's guts. Secondly, safe for the use of the small mammal with no significant adverse effects. Thirdly, this vaccine can be used for different species of small mammals and is effective towards different *Borrelia* bacteria.

On a smaller scale, an effective insect repellent that specifically targets ticks could also be developed and commercially sold. This could also reduce the chances of tick bites whilst in tick infested areas and potentially reduce the number of infections.

Conclusion

In conclusion, Lyme disease is a major emerging disease that is caused by *Borrelia burgdorferi* which is a bacteria that is transmitted from natural reservoirs to ticks and then to humans during tick blood meals. The bacteria is able to evade the immune system and cause this disease. This disease can be fatal if not treated during early stages with antibiotics and there is currently no vaccines to prevent this disease beforehand. This disease has not been viewed as seriously by the general public and has not attracted the pharmaceutical interest than other diseases such as HIV/AIDS and Tuberculosis but the consequences of this disease can be just as great (Daniel J C *et al.*, 2014). Moreover, human's knowledge on Lyme disease is also less compared to the other diseases which greatly prohibits humans to develop suitable prevention methods towards this disease.

<u>Reference</u>

About Ticks - Lyme Disease Association of Australia. (2016). *Lymedisease.org.au.* Retrieved 19 May 2016, from http://www.lymedisease.org.au/about-lyme-disease/about-ticks/

Borrelia burgdorferi (Lyme disease). (2016). *YouTube*. Retrieved 16 May 2016, from https://www.youtube.com/watch?v=EBFzGaoHMhQ

Borrelia burgdorferi and Lyme Disease - MicrobeWiki. (2016). *Microbewiki.kenyon.edu.* Retrieved 13 May 2016, from https://microbewiki.kenyon.edu/index.php/Borrelia_*burgdorferi*_and_Lyme_Disease

Break Through in Lyme Disease. (2016). *Holtorf Medical Group*. Retrieved 13 May 2016, from https://www.holtorfmed.com/breakthrough-in-lyme-disease/

Cameron, D., Johnson, L., & Maloney, E. (2014). Evidence assessments and guideline recommendations in Lyme disease: the clinical management of known tick bites, erythema migrans rashes and persistent disease. *Expert Review Of Anti-Infective Therapy, 12*(9), 1103-1135. http://dx.doi.org/10.1586/14787210.2014.940900

Critical Needs and Gaps in Understanding Prevention, Amelioration, and Resolution of Lyme and Other Tick-Borne Diseases. (2011). http://dx.doi.org/10.17226/13134

Dangers Of Tick Borne Lyme Disease. (2016). *YouTube*. Retrieved 15 May 2016, from https://www.youtube.com/watch?v=Fd-W_dyd204

Data and Statistics| Lyme Disease | CDC. (2016). *Cdc.gov*. Retrieved 11 May 2016, from http://www.cdc.gov/lyme/stats/

Epidemiology of Lyme disease. (2016). *Uptodate.com*. Retrieved 10 May 2016, from http://www.uptodate.com/contents/epidemiology-of-lyme-disease?source=see_link

Garcia-Monco JC, e. (2016). *Adherence of the Lyme disease spirochete to glial cells and cells of glial origin. - PubMed - NCBI. Ncbi.nlm.nih.gov*. Retrieved 14 May 2016, from http://www.ncbi.nlm.nih.gov/pubmed/2760500

Gofton, A., Doggett, S., Ratchford, A., Oskam, C., Paparini, A., Ryan, U., & Irwin, P. (2015). Bacterial Profiling Reveals Novel "Ca. Neoehrlichia", Ehrlichia, and Anaplasma Species in Australian Human-Biting Ticks. *PLOS ONE, 10*(12), e0145449. http://dx.doi.org/10.1371/journal.pone.0145449

Growing evidence of an emerging tick-borne disease that causes a Lyme like illness for many Australian patients – Parliament of Australia. (2016). *Aph.gov.au*. Retrieved 13 May 2016, from http://www.aph.gov.au/Parliamentary_Business/Committees/Senate/Community_ Affairs/Lyme-like_Illness

History of Lyme Disease | Bay Area Lyme Foundation. (2016). *Bay Area Lyme Foundation*. Retrieved 22 May 2016, from http://www.bayarealyme.org/about-lyme/history-lyme-disease/

How Does Lyme Disease Evade the Immune System. (2016). *Holtorf Medical Group*. Retrieved 11 May 2016, from https://www.holtorfmed.com/lyme-disease-evade-immune-system/

How Infection Works, How Pathogens Make Us Sick — The National Academies. (2016).*Needtoknow.nas.edu*. Retrieved 11 May 2016, from http://needtoknow.nas.edu/id/infection/how-pathogens-make-us-sick/

How Lyme Disease and Its Treatments Work. (2016). *YouTube*. Retrieved 13 May 2016, from https://www.youtube.com/watch?v=RTiWfyrNBwA

Klinghardt: Biological treatment of Lyme disease. (2016). *Klinghardtacademy.com*. Retrieved 11 May 2016, from http://www.klinghardtacademy.com/Protocols/Klinghardt-Biological-treatment-of-Lyme-disease.html

Kovacs-Simon, A., Titball, R., & Michell, S. (2010). Lipoproteins of Bacterial Pathogens. *Infection And Immunity, 79*(2), 548-561. http://dx.doi.org/10.1128/iai.00682-10

Kurtenbach K, e. (2016). *Vaccination of natural reservoir hosts with recombinant lipidated OspA induces a transmission-blocking immunity against Lyme disease spirochaetes a... - PubMed - NCBI. Ncbi.nlm.nih.gov*. Retrieved 11 May 2016, from http://www.ncbi.nlm.nih.gov/pubmed/9364698

Late Stage Lyme Disease - Lyme Disease Association of Australia. (2016). *Lymedisease.org.au*. Retrieved 9 May 2016, from http://www.lymedisease.org.au/about-lyme-disease/late-stage-lyme-disease/

Lyme Disease. (2016). *Healthline*. Retrieved 15 May 2016, from http://www.healthline.com/health/lyme-disease#Description1

Lyme Disease. (2016). *Niaid.nih.gov*. Retrieved 10 May 2016, from https://www.niaid.nih.gov/topics/lymedisease/Pages/lymeDisease.aspx

Lyme Disease -- History and Current Controversies. (2016). *YouTube*. Retrieved 11 May 2016, from https://www.youtube.com/watch?v=yC_lEERJXSA

Lyme Disease | Patients Receive Last Chance at Lyme Disease Treatment. (2016). *Envita Medical Center*. Retrieved 15 May 2016, from https://www.envita.com/conditions/lyme-disease

Lyme disease graphs | Lyme Disease | CDC. (2016). *Cdc.gov*. Retrieved 13 May 2016, from http://www.cdc.gov/lyme/stats/graphs.html

Lyme disease transmission. (2016). *Cdc.gov*. Retrieved 5 May 2016, from http://www.cdc.gov/lyme/transmission/

Lyme disease treatment. (2016). *Uptodate.com*. Retrieved 14 May 2016, from http://www.uptodate.com/contents/lyme-disease-treatment-beyond-the-basics

LYME DISEASE: Why don't we just eradicate ticks?. (2016). *Nhregister.com*. Retrieved 15 May 2016, from http://www.nhregister.com/article/NH/20120621/NEWS/306219964

Lyme Disease| Lyme Disease | CDC. (2016). *Cdc.gov*. Retrieved 17 May 2016, from http://www.cdc.gov/lyme/

Many patients with chronic Lyme disease are profoundly debilitated.. (2016). *LymeDisease.org*. Retrieved 7 May 2016, from https://www.lymedisease.org/lyme-basics/lyme-disease/chronic-lyme/

Marques, A. (2008). Chronic Lyme Disease: A Review. *Infectious Disease Clinics Of North America,22*(2), 341-360. http://dx.doi.org/10.1016/j.idc.2007.12.011

Microbiology of Lyme disease. (2016). *Uptodate.com*. Retrieved 11 May 2016, from http://www.uptodate.com/contents/microbiology-of-lyme-disease

Pathogen. (2016). *ScienceDaily*. Retrieved 14 May 2016, from https://www.sciencedaily.com/terms/pathogen.htm

Poland, G. (2011). Vaccines against Lyme Disease: What Happened and What Lessons Can We Learn?.*Clinical Infectious Diseases*, *52*(Supplement 3), s253-s258. http://dx.doi.org/10.1093/cid/ciq116

Post-Treatment Lyme Disease Syndrome| Lyme Disease | CDC. (2016). *Cdc.gov*. Retrieved 15 May 2016, from http://www.cdc.gov/lyme/postlds/

Science, I. (2011). Federal Funding of Tick-Borne Diseases. *National Academies Press (US)*. Retrieved from http://www.ncbi.nlm.nih.gov/books/NBK57017/

Science, L. (2013). *Weird Way Lyme Disease Bugs Avoid Immune System. Live Science*. Retrieved 14 May 2016, from http://www.livescience.com/28120-lyme-disease-manganese.html

Signs and Symptoms | Lyme Disease | CDC. (2016). *Cdc.gov*. Retrieved 10 May 2016, from http://www.cdc.gov/lyme/signs_symptoms/

Stages of Lyme Disease-Topic Overview. (2016). *WebMD*. Retrieved 12 May 2016, from http://www.webmd.com/arthritis/tc/stages-of-lyme-disease-topic-overview

Stricker, R. & Johnson, L. (2014). Lyme Disease: Call for a "Manhattan Project" to Combat the Epidemic. *Plos Pathog, 10*(1), e1003796. http://dx.doi.org/10.1371/journal.ppat.1003796

The Biology of Lyme Disease: An Expert's Perspective (Part 1 of 3). (2016). *YouTube*. Retrieved 16 May 2016, from https://www.youtube.com/watch?v=r8tESJVvM88

The History of the Lyme Disease Vaccine — History of Vaccines. (2016). *Historyofvaccines.org*. Retrieved 18 May 2016, from http://www.historyofvaccines.org/content/articles/history-lyme-disease-vaccine

Treatment| Lyme Disease | CDC. (2016). *Cdc.gov*. Retrieved 15 May 2016, from http://www.cdc.gov/lyme/treatment/

Vaccines for Lyme Disease - Past, Present, and Future. (2016). *YouTube*. Retrieved 16 May 2016, from https://www.youtube.com/watch?v=vzSih6IHwYA

Source Analysis

Title	Chronic Lyme Disease: An appraisal
Author(s)	Adriana Marques
Nature of Publication	Scientific Article
Date of Publication	2008
Publisher	National Center for Biotechnology Information (NCBI)
Bias of Publisher	There should not be any bias as NCBI is a part of the United States National Library of Medicine and is also a branch of National Institution of Health which is a part of the Department of Health and Human Service. Thus, NCBI is a government institution and the information given should be reliable and not bias.
Purpose of Publication	The purpose of the publication is to provide to the public, information on chronic Lyme disease and its dangers.
Qualifications of author(s)	The author is the chief of the Clinical Studies Unit, Laboratory of Clinical Infectious Disease and mainly researches Lyme disease and Herpes simplex virus infections.
Bias of author	The author's bias should be limited as she mainly researches on Lyme disease and she works in a laboratory which mainly focuses on areas such as Lyme disease. Thus, her bias should to a minimum
Assessment of reliability, authenticity and bias of publication	The publisher is a government owned institution and the author is a professional in Lyme disease and works for a science laboratory. Thus, this source should be reliable.

Title	Vaccines against Lyme Disease: What Happened and What Lessons Can We Learn?
Author(s)	Gregory A. Poland
Nature of Publication	Scientific Journal
Date of Publication	2011
Publisher	Oxford Journals/Oxford University Press
Bias of Publisher	Bias should be limited as the Oxford Journal is from a famous University and is also the largest and second oldest university press in the world, therefore, its articles are most likely reliable.
Purpose of Publication	Inform the public about the events that caused the old Lyme disease vaccine to withdrawn and what we need in the future.
Qualifications of author(s)	The author studies the immunogenetics of vaccine responses in adults and children and thus, is an expert in the areas of the development of vaccines.
Bias of author	The author's bias is limited as well as he's research are continuously funded by large government institutions such as the National Institution of Health
Assessment of reliability, authenticity and bias of publication	Oxford Journals or the Oxford University Press is a department of the University of Oxford and the author is a professional in these areas and he is continuously researching with the Vaccine Research Group of the Mayo Clinical. Thus this source has high reliability.

Title	Lyme Disease: Call for a "Manhattan Project" to Combat the Epidemic
Author(s)	Raphael B. Stricker and Lorraine Johnson
Nature of Publication	Scientific Article
Date of Publication	2014
Publisher	PLOS Pathogens
Bias of Publisher	This publisher should not be bias as this is a medical journal and is peer reviewed publication. Thus, this publisher should not be bias since the publication is reviewed by other scientists.
Purpose of Publication	The purpose of the publication is to call on scientists to combat Lyme disease
Qualifications of author(s)	The author, Raphael B. Stricker is a haematologist which studies the diagnosis, prevention and treatment of diseases. Lorraine Johnson is also the Chief Executive Officer of LymeDisease.org
Bias of author	The author's bias is limited as one author is a professional and the other is the chief officer at a healthcare institution.
Assessment of reliability, authenticity and bias of publication	The source is authentic and reliable as both the publication which is a professional scientific journal and both authors have qualifications that makes them experts at these areas.

Title	Vaccine for Lyme Disease – Past, Present and Future
Author(s)	Stanley Plotkin, Rich Marconi, Robert Aronowitz, Nina Wressnigg, Larry Madoff
Nature of Publication	Scientific Webinar
Date of Publication	2015
Publisher	Centers for Disease Control and Prevention
Bias of Publisher	The publisher is not likely to be bias as the publisher, Centers for Disease Control and Prevention, is a public health institute and a federal agency under the United States Department of Health and Human Service. This is a government institution and thus the bias should be very limited.
Purpose of Publication	The purpose of the publication is to inform and point out to the public on the seriousness of Lyme disease and the need for a new vaccine.
Qualifications of author(s)	The authors are all qualified professionals in their areas of studies such as the first author, Stanley Plotkin is an American physician who works as an adviser at a pharmaceutical firm.
Bias of author	The author's bias is also limited as the authors are professionals and thus the information they provide should not be bias. However, there is some controversy topics and their own opinions may be provided.
Assessment of reliability, authenticity and bias of publication	As both the publication and the authors are authorities in this area and they have limited bias, the information provided should be reliable.

Title	Evidence Assessments and Guideline Recommendations in Lyme Disease: The Clinical Management of Known Tick Bites, Erythemia Migrans Rashes and Persistent Disease.
Author(s)	Daniel J Cameron, Lorraine B Johnson and Elizabeth L Maloney
Nature of Publication	Scientific Journal
Date of Publication	2014
Publisher	Taylor and Francis
Bias of Publisher	The publisher is most likely not bias as this publisher publishes academic journals and its published works are peer reviewed.
Purpose of Publication	The purpose of publication is to create recommendation and assess the clinical management of Lyme disease
Qualifications of author(s)	These authorities are experts in the area of treating Lyme disease. As an example, Dr Daniel J Cameron is a national leading expert in diagnosing Lyme disease
Bias of author	The bias should be minimal from the author since these authors are well known professionals in this area and putting bias will make their journals to be unprofessional.
Assessment of reliability, authenticity and bias of publication	Both the publication and the authors do not have bias and the publication have a peer reviewing system, the authors are already well known experts. Thus, this source is reliable.